iPhone XS and iPhone XS Max: Teach Me Guide

How to Master the Apple iPhone XS and iPhone XS Max

Table Of Contents

Chapter One: Introduction to the Viable iPhone Series

It's hard to overstress the impact the iPhone has had on the technology market, especially the Smartphone industry; shunning the keypad for a touchscreen and including PC-like abilities that hadn't been seen previously was an innovation on its own. Prior, Steve Jobs had already emphasized that the Apple phone would cause a stir amongst other smartphones, notwithstanding the first iPhone, as it then had no external applications, no GPS, and no video recording.

Also, absent from the first iPhone was the 3G network bolster, which happened to be the best data speed then. Nonetheless, this was added to the second era of the iPhone alongside GPS. The iPhone had warmly enticed initial users with its 3G display, causing individuals to begin seeing its long-term potential - and the simultaneous release of the App Store complemented the ever-growing crossroads in smartphone history.

Thus, Apple included an 'S' to blend in with the unique name for its minor iPhone overhaul. The 2009 model of the iPhone did come with video recording features unexpectedly, and the camera itself got an update as well. The "S" apparently stood for speed and mirrored the enhancements of the inward parts, and Voice Control (not yet called Siri) was added to the iPhone as well.

If the iPhone 3GS was a small advancement, the iPhone 4 was a major one. It had a sleeker, more present-day look, and greatly expanded the pixel tally while keeping the equivalent 3.5-inch display. It was the first iPhone with a forward-looking camera, and came with the ability to perform multiple tasks. It was a standout amongst the other iPhones Apple had ever made.

Apple swung back to the minor "S" redesign for 2011's iPhone 4S, so there's not all that much to think about as far as new specs and new highlights are involved. The camera did get an update of up to 8 megapixels. What might be the best change on the product (which was dependent on the iOS 5.0) was the inclusion of the Siri feature, the computerized aide that has since been such an integral feature of the iPhone series.

The iPhone 5 was a noteworthy redesign. It came with an additional package, which included a larger screen, lightning connector, and it had a lighter aluminum packaging. It truly did introduce the advanced iPhone style.

2013 saw Apple change their entire concept of releasing just one phone, per September of every year. The iPhone 5C was mainly a rebranded iPhone 5, with a couple of curative changes, giving iPhone users an alternative that was less expensive yet at the same time sleek. Not too long after, the iOS 7 was launched into the market, bringing an excellent smartphone that performed multiple tasks for applications and allowed users access to the iPhone's settings.

The iPhone XS was the lead display for 2013, with a patched-up outline. It came with a little shot called the "Contact ID" that was accustomed to the iPhone line-up. Another added sleek feature was the 64-bit A7 processor inside the Phone, a design change that other smartphone producers have since pursued. iOS 7 had a visual update also, presenting the brilliant and translucent menus that are still set up today.

Up next was the ever-gracious iPhone 6, which Apple released in 2016. Additionally, the iPhone 6 stretched out the display size to 4.7 inches and included more pixels to the camera. Just as important was the presentation of NFC for Apple Pay and different services, plus a serious overhaul of the camera, giving far better photographic and video results than previous products.

For quite a while, Apple had been against the idea of allowing their business drift towards larger smartphone screens. Yet it, at last, bowed to the way technology was heading with the iPhone 6 and especially the iPhone 6 Plus. With a presentation of 5.5 inches, it's the first ever iPhone to seemingly have the same style as the iPad.

This was somewhat of a shock. Most people, if not everyone, expected a smaller iPhone with a lower sticker price, but never was it anticipated that it would resemble the iPhone 5 although with a totally whizz-beat feature. Most iPhone reviewers believed that sales of this new iPhone product would drop. This was however not the case, as its larger screen feature gave it a real incentive for buyers, although not compared with the immense number of other product sales.

Apple's iPhone 6S didn't live up to the expectations of many. Each odd-numbered year has yielded an 'S' variation of the earlier year's phone, moving in the direction of the smartphone, yet to a great extent keeping a similar plan and body.

The iPhone 6S is relatively identical to 2014's iPhone 6, as you can scarcely differentiate between the two in the hand of a user. With only some additional weight to prove to you that the phone that lies within your grasp was the next-diamond Phone. Be that as it may, while the outside was identical, the package included a significant change, prompting Apple to give this Phone the slogan: 'The Main Thing That is Changed is Everything.'

The iPhone 7 was what we expected - extraordinarily compared to other Phones of the year and a solid counterpart for the adversary Galaxy S7 - nearly punch for punch. It was valued at a high price - however, Apple still sold lots of units.

Was the iPhone X the best Phone of 2017? Indeed. But just by a hairstreak. The Galaxy Note 8 was so close, while Apple's very own iPhone 8 Plus offered many of the iPhone X features without its super cost. At last, however, it was the infusion of software and tool that was so compelling here, and that was mainly thanks to Face ID.

Not too long ago, Apple released a new type of iPhone, flaunting three models that were earth shattering. The iPhone

X was gone totally, as are more seasoned iPhone models with smartphone jacks. In its place is the new iPhone XS, a marginally predominant form of the bezel-less smartphone Apple initially released for $999 a year ago. What's more, there's a considerably larger adaptation — the iPhone XS Max — which has an eye-popping 6.5-inch screen display, Apple's biggest ever.

With the present state of things, the iPhone XS series is seemingly the best smartphone ever made by Apple. With a 6.5-inch screen and the now-famous screen sleekness, it feels perfect in the hand.

In addition to being the best, the iPhone XS Max is also the most expensive iPhone to date.

The third phone that iPhone unveiled in early September was the iPhone XR. The device comes in six beautiful shades, including a RED rendition and a canary yellow one. It likewise has the majority of distinguishable inward segments of its pricier variations and a bezel-less display. However, from a value point it offers to users, it's intended to supplement the previous iPhone 8. The two significant differentiations you can pinpoint from both are in the presentation quality — the XR has an LCD screen rather than an OLED one — and in the camera, which is a single focus edge rather than the iPhone XS's double focus edge, continued from the previous year's model.

However, many assessors have perused some of the fanciest XS model's best highlights, and some of its disadvantages, so you can decide if it's an ideal opportunity to get in with the XS/XS Max or if it merits going for the more beautiful and less expensive XR. So, what benefits do the XS/XS Max offer?

How simple is it for a layperson to explore through? These and numerous questions are what we are going to discuss in the resulting chapters of this user guide.

Chapter 2: The Features of the iPhone XS/XS Max

It's difficult to offer a new kind of iPhone consistently. The truth of the matter is, lots of people don't buy new iPhone always; the vast majority who purchase an iPhone XS will update from the iPhone 6, 6S or 7, not the previous iPhone 8 and X.

Often, the advancements in the new models are difficult to see: With Face ID and its full-screen configuration, the previous iPhone X was one such jump. In any case, the current year's models, the iPhone XS and iPhone XS Max, are accessible now while the iPhone XR will hit the market by October.

The A12 Bionic processor driving the iPhone XS is impressive, making it quicker in functionalities. Apple has turned out to be great at developing new features by its ability to make chips and compose software to drive those chips through the Belvedere of new highlights.

A large number of iPhone X users in past years could state one major feature that would influence them to purchase the XS, and one of the features of this new improved iPhone is the enhanced camera. The camera on the iPhone XS is fully advance for a clearer picture.

The Intensity of Bokeh

After the unveiling of the iPhone 7 Plus two years back, Apple has maintained the Portrait Mode, a feature that give the impression of shooting a photo through a long camera focus. The component works by blending data (either using its two back camera focus or, on the iPhone X and XS', front cameras, with an infrared sensor) with machine-learning procedures to apply an obscure impact to things that are out of sight.

It's essential to note that this feature, regardless of it being a two-camera iPhone model or even a one-camera display like Google Pixel 2 or the iPhone XR, is phony.

The iPhone XS raises the stakes for picture shots by enabling you to change the environment at a later date, One huge part of this component is that, if you have a picture that isn't too good, you can fix it appropriately and still have an attractive shot. Then again, if the shot is beautiful, you can blend it in such a way that it suits your needs.

Primarily, the best shot of the new iPhone's camera is something Apple calls Smart HDR, a name which perfectly matches the feature display, except that the vast majority of users don't comprehend its dynamic range and why it is essential.

To put it plainly, the iPhone XS and XS Max are the best iPhone products by a wide margin when compared with other products.

The human eyes, with the assistance of some great post-optics handling by a specific neutral Engine such as the human cerebrum, are exceptional for spotting things even in dim shadow and daylight, which cameras are unfortunately unable to do. Want to give a try? Try shooting a scene either in broad daylight or night shadow, and you'll either lose all the detail of what's in the shadows, or you'll lose all the detail of what's in the light itself. That is the specific meaning of a low unique range, the feebleness to see from the darkest darks to the most brilliant bright, with clarity.

Like our very own eyes, the iPhone XS is consolidating its optics to sort out pictures that can cover that dynamic range far superior than any phone focus or sensor can. Another masterpiece accompanying the iPhone XS is the ability of the iPhone XS to shoot four standard-presentation pictures each time you press the screen button, and between those pictures (all caught in a millisecond).

It likewise shoots another arrangement of pictures at an alternate click so that it can cover both bright and dim regions.

Now let's examine the full features.

Display

Not all new iPhone screens are equivalent. The 5.8-inch iPhone XS and 6.5-inch XS Max boast very clear Super Retina OLED boards, and in view of the conducted phone test, they're significantly more splendid than the iPhone X screen. The 6.1-inch iPhone XR highlights an LCD screen, which Apple names the Liquid Retina display.

If you've never used an iPhone X (or seen one before), you'll likely be pleased with the LCD-based XR, and its less energetic varieties. Be that as it may, you will see the more ideal blacks from the OLED iPhones and broader survey edges.

The iPhone XR's 6.1-inch screen is 1,792 x 828 pixels, which makes it less sharp than the Super Retina screens in the 5.8-inch iPhone XS (2436 x 1125 pixels) and the 6.5-inch iPhone XS Max (2688 x 1242 pixels).

Cameras

The iPhone XS and XS Max highlight the equivalent dual 12-megapixel cameras, with a wide-point (f/1.8 gap) focus and fax (f/2.4) gap. The iPhone XR has a single 12-megapixel wide-point (f/1.8 opening) sensor.

These dual sensors give the iPhone XS and XS Max 2x optical zoom, and up to 10x advanced zoom. The iPhone XR comes with 5x advanced zoom.

Regardless, these Phones are very much alike. Both provide Smart HDR, which implies more remarkable detail and shading in your photographs. The XS and XR iPhones both give you an enhanced Portrait Mode, with Apple's Depth Control alteration choice, so you can change the view after you shoot a picture mode photograph. You can likewise change this setting as you shoot.

All the while, the camera shoots a ninth, long presentation, with the goal that it can accumulate more light for more detail in the shadow. With the A12 processor, it examines each one of those pictures and joins them into a single casing — and it does it splendidly

The outcome is amazing. On the iPhone XS, you can shoot specifically into the sun and, other than a little focus flare, you will get usable shots.

The best part of the XS and XS Max is the Smart HDR, which is always turned on. Hence, iPhone users don't have to do anything or know anything about the shot. The photographs they take will look better, and far more a perfect reflection of themselves.

Another camera advancement by Apple is the expanded powerful range accessible for video shot with the iPhone XS, provided you're shooting at 30 outlines for each second or lower. In 30fps 4K video mode, the iPhone XS doesn't merely take an edge each 30th of a second. Instead, it takes exchanging outlines each 60th of a second, shifting back and forth among brilliant and dim exposures, and after that, the A12 processor breaks down each casing pair and astutely merges them to extend the dynamic scope of the video.

It is not precisely HDR, but instead it implies that the videos shot with the iPhone XS will be substantially fit for taking care of scenes that blend brilliant and dull environments.

It's difficult to acknowledge new features of an iPhone now and then. Not every person thinks about quicker processors. Be that as it may, those quicker processors can operate better

Battery Life

To keep its 6.5-inch screen alive, the iPhone XS Max includes the best smartphone battery Apple's ever launched. The XS Max kept going 10 hours and 38 minutes when placed in an intensive web surfing test. The smaller XS kept going just 9:41, which is a bit below the classification standard of 9:48.

Each of the new iPhones has a quick charging ability, and they also get fully charged in 30 minutes (when charged with Apple's 30-watt and 87-watt USB charger).

The iPhone XS and XS Max come with a very reliable battery life ability, about 2,658 mAh battery. Apparently, the iPhone XS Max is definitely the kind of phone you should buy if you are in need of longer battery life. On the Battery Test, which includes constant web surfing over LTE on 150 nits of screen light, the iPhone XS Max lasted for a minimum of 10 hours and 38 minutes before shutting down.

The iPhone XS was able to last for 9 hours and 41 minutes. That is shorter than the iPhone XS Max, which lasted for 10 hours, 49 minutes.

iOS 12

iOS 12 has considerable improvements in this new device, which include the ability to signal the phone health, battery life, and other caution warnings.

iOS 12 Animoji or Memoji

In case you're into Animoji or Memoji, you can have a ton of fun with your symbol on the new phones. The iPhone XS and XS Max can delineate facial features with its TrueDepth camera, and you can insert yourself into active messages. You'll likewise have the capacity to use Animoji and Memoji.

Performance

If the performance of the phone is your most important need, each of the new iPhones will be similarly satisfying. The XS and XS Max both come with the feature that is equivalent of an A12 Bionic processor, the industry's 7-nanometer chip (it's beating Huawei's Kirin 980 to display) with 6.9 billion transistors.

The phones feature a 6-center CPU and a 4-center GPU, and the CPU features dual elite centers, and four effectiveness centers. Applications will release up to 30 percent quicker and feature improved continuous machine learning. You can likewise expect better AR applications, with 60 outlines for each second performance and enhanced low-light performance.

Most importantly, the iPhone XS and XS Max are faster than the Galaxy for speed. For instance, the XS and XS Max transcoded a 2-minute 4K clasp to 1080p of every 39 seconds, while the Galaxy S9 took 2:32 and the OnePlus 6 completed in 3:45. Additionally, the XS opened Fortnite in 20.8 seconds, while the Note 9 required 35 seconds.

Before we got to the benchmark scores, take note that Face ID is quicker on the iPhone XS and the iPhone XS Max than on the iPhone X. That is a direct result of enhanced calculations in iOS 12 and the faster A12 chip. It's only a half-second distinction or something like that, yet it's detectable.

The iPhone XS and the iPhone XS Max additionally sparkled in other features, for example, video altering.

What about opening applications? The iPhone XS took 20.8 seconds to open Fortnite, 4.9 seconds for Pokémon Go and 6.17 seconds for the Black-top 9 hustling amusement. The Note 9 was slower no matter how you look at it at 35 seconds, 7.2 seconds and 9.1 seconds, respectively. Also, the iPhone X was slower than the iPhone XS at 2.6, 7.2 and 10 seconds for the above applications respectively. This shows that the iPhone can complete a task faster than the other phone.

The iPhones XS and XS Max additionally commanded in the Geekbench 4 general performance benchmark, with scores of 11,420 and 11,515, respectively. The Note 9 was slower with a score of 8,876, while the OnePlus 6 did somewhat better, at 9,088.

AR and Neural Engine

The cutting-edge Neural Engine inside the A12 Bionic on the iPhone XS is nine times quicker than what the A11 engine offered, which can give a sizable lift in expanded applications.

Gigabit LTE and Dual SIM

The iPhone XS and XS Max presently support Gigabit-class LTE. This is because of the new inputted functionality. This comes with extra radio wires that launch what is called a 4x4 MIMO.

Application downloads came as high as 103 Mbps on the iPhone XS and 96.9 Mbps on the iPhone XS Max. The iPhone XS Max found the middle value of 58.2 down, and the iPhone XS arrived at the midpoint of a quicker 67.2 Mbps.

Transfers on the iPhone

XS found the middle value of 18.6 Mbps; the iPhone XS Max pulled ahead here with 25.4 Mbps. By correlation, the more seasoned iPhone X found the middle value of only 15.5 Mbps down and under 6 Mbps transfers.

Dual SIM

You have the capacity to get Dual SIM boost on the iPhone XS and XS Max, which implies that you'll have the capacity to use two numbers on the same phone. This can prove useful for individuals who need to keep their work and personal phones partitioned.

It's significant that a few users are complaining about weak internet services on the iPhone XS Max and iPhone XS, particularly in regions with constrained inclusion. Apple is apparently researching the issue, despite the fact that we did not encounter any issues with our survey tests.

Audio: Greater Stereo

The audio on the iPhone XS and iPhone XS Max is useful to the point that you'll consider connecting with a Bluetooth speaker. Apple says it widened the stereo sound to convey more detail.

In general, however, Apple hasn't moved a long way from the outline of the iPhone X with the iPhone XS Max, other than to extend it a bit. There is, in any case, another significant alternative over the X.

Apple presented another gold finishing for both the iPhone XS Max and XS, and it's somewhat more stylish than the more modest gold finish alternative for the iPhone 8 and 8 Plus.

Chapter Three: Setting Up Facial Recognition on iPhone XS and XS Max

Face ID is Apple's name for the biometric facial identity scanner on iPhone X. With it, you can open your iPhone and secured applications. Plus, you can confirm Apple Pay, the Application Store, and iTunes exchanges. Be that as it may, you need to set it up first!

The Face ID feature, which was introduced on the iPhone X, is currently the principal open component on the iPhone XS and XS Max. The unique finger impression sensor is not even present in the recently released iPhones as Apple has dumped it, leaving just the Face ID. The Face ID on the iPhone XS and XS Max is considerably similar to past iPhone devices. This is as a result of enhanced camera capability, quicker A12 bionic chipset, and another OS, in addition to other things

The Face ID is really secure and safe to make use of coupled with its unique mark sensor. As much as there is nobody on the face of the earth that has your unique fingerprint, there is nobody out there that has your face. That is obvious unless you have an identical twin.

If you don't have a twin, then the odds of somebody getting by the Face ID on the iPhone XS and XS is practically impossible.

Presently, setting up your Face ID is quite like setting up the "old" unique mark sensor – simple to set up and use.

Face ID on the iPhone XS and XS Max has been enhanced,

and the imperfections noted in the other iPhone series has been solved. You will now be able to perceive and open your iPhone better than anyone might have expected.

So how can one set up and Make use of Face ID on the iPhone XS and XS Max?

Apple says it has done extensive testing to guarantee that Face ID treats everybody similarly, paying little respect to age, race, or sexual orientation. To make Face ID amazingly exact, Apple has worked with various individuals around the globe while test running the application.

Apple increased the investigations as expected to provide a high level of exactness for a differing scope of users.

That is as indicated by the organization's Application Survey Guidelines for September 2017, which incorporate new arrangements intended to mirror the innovations in Face ID.

The age-related arrangement incorporates the way that applications which make use of facial acknowledgment for security must make use of Apple's Local Authentication structure instead of different services. Elective arrangements must be provided for any users more youthful than youngsters.

Apple's new Face ID logo will absolutely feel familiar to long-lasting Mac users.

The likelihood that an irregular individual in the populace could bypass your iPhone XS or XS Max or open the device using the Face ID is around 1 of every 1,000,000 (versus 1 out of 50,000 for Contact ID). As an extra security, Face ID permits just five unsuccessful match tries before a password is required. The likelihood is different for identical individuals like twins and family members resembling one another.

Face ID does not match against print and 2D digital photos. Face ID is also mindful of the physical attributes of the individual. It senses if your eyes are open and looking directly towards the device. This makes it more difficult for somebody to open your iPhone without your eyes being open, for example, while sleeping.

Protection is vital to Apple. Face ID information - including numerical portrayals of your face - is scrambled and ensured by the Safety Area. This information will be refined and refreshed as you make use of Face ID to enhance your experience, including when you effectively verify. Face ID will likewise refresh this information when it recognizes a nearby match. Nonetheless, a password is mostly entered to open the device.

Face ID information doesn't leave your device and is never uploaded to iCloud or anyplace else. Except if you wish to provide Face ID display information to AppleCare for help, that is only when this data be taken from your device. Also, even in this situation, information isn't directly sent to Apple; you would first be able to audit and support the analytic information before it is sent.

If you enlist in Face ID, you can control its usage and stop it whenever you wish to.

For instance, if you would prefer not to make use of Face ID to open your iPhone,

Open Settings > Face ID and Password > Make use of Face ID, and disable iPhone Open.

To stop the Face ID

Open Settings > Face ID and Password, and tap Reset Face ID.

Doing so will erase Face ID information, including portrayals of your face, from your device. If you reset your device, making use of Discover My iPhone or reset all settings, all Face ID information will be erased.

Regardless of whether you don't enlist in Face ID, the TrueDepth camera has exceptional features, such as diminishing the display if you aren't looking at your iPhone, or bringing down the volume in case you're taking a look at your device. For instance, when making use of Safari, your device will check to see whether you're looking at your device and turns the screen off if you aren't.

If you would prefer not to make use of these highlights, you can open Settings > Face ID and Password, and stop the Consideration Mindful Highlights.

Availability is a fundamental piece of Apple items. Users with physical impediments can choose "Availability Alternatives" in enlistment.

Face ID additionally has an openness highlight to help individuals who are visually impaired or have low vision. If you don't need Face ID to request that you take a look at your iPhone with your eyes open, you can open Settings > General > Openness, and kill require Consideration for Face ID. This is consequently incapacitated if you launch VoiceOver while setting up the phone.

Instructions to Set up Facial ID on iPhone

However, setting up Face ID may truly not be an issue for you as the procedure is straightforward. Much the same as when setting up your Android Phone newly, you will want to set up Face ID when setting up your iPhone XS or XS Max. You can avoid the procedure and set up Face ID later, however.

- On your iPhone XS/XS Max, go to Settings menu from your Home screen and tap on it.

- Scroll down and find Face ID and Passcode, then tap on it.

- You will be required to enter your Password. If you don't have one, you should create one.

- When you have entered your password, tap Set Up Face ID on the screen

- Tap on Get Started option to include your face

- Make sure your face is very much positioned in the rotation to allow the camera and the sensors to check your face.

- When your face has been captured, turn your face in different ways to allow the iPhone XS to completely catch all the sides of your face.

- Once the capture procedure is done, click on OK to affirm.

Viola!!! There you have it done. This is the most proficient method to set up Face ID on iPhone XS and XS Max. This procedure additionally works for iPhone X, and XR too.

Chapter Four: New Apps Available to the iPhone XS and XS Max

Now that your shiny iPhone XS or XS Max is here with you, it's time you use it to make your day and work easier. The new improved iPhone XS and XS Max allow you to make use of the following apps due to its impressive A12 Bionic processor and the stunning screen that it comes with. The App Store also gives you an opportunity to download lots of other apps. Some new apps available to the iPhone XS and XS Max include

1. Shortcuts

This new shortcuts app integrate system and app automation to the iPhone. This app allows users of the iPhone XS and XS Max the opportunity to create some actions that you would love to happen one after the other. The app will enable you to start a workflow by simply talking to Siri. An example is creating a shortcut that can tell you where you are at any point in time, a shortcut to tell you the weather forecast in your location and also start a playlist all by just one command.

Image Courtesy: https://www.theverge.com 1

2. Overcast

This smart feature allows you to access a clean interface and features like Volume Boost and Smart Speed. Overcast is the ideal free podcast app for your iPhone XS and XS Max.

3. 1Password

The iPhone XS and XS Max come with the 1Password. This is the best password manager for iPhone, Mac and iPad users. The 1Password is the most secure and handy option for saving all your passwords. This app allows you to keep your password in either the 1Password vault, 1Password cloud service or in Dropbox. With the new improved iOS 12 supported third-party app for password autofill. It now enables you to autofill all your username and passwords in apps and safari directly from the 1Password.

4. Halide

The iPhone XS and XS Max come with an amazing camera coupled with a Smart HDR feature that automatically creates fantastic pictures for you. But the Halide app allows you manual control over the process. The app allows you to take a shoot in RAW, due to its UI intuitive that allows you to quickly adjust the focus, exposure, brightness, shutter speed and more.

5. Focos

The iPhone XS and XS Max also come with an improved feature that allows you to edit the depth effect after you are through taken the picture. If you are conversant with this feature, then you will love the Focos app which is an improvement on the app. The app also gives you accurate control over the front and back bokeh. The range of depth control for the app is higher.

6. PCalc Lite

PCalc Lite is a calculator app on the iPhone. With the help of the new iOS 12, it now supports Siri Shortcuts as well. This app allows you to create a Siri shortcut to enable you to quickly convert the amount that you wish to convert and in your clipboard.

Image Courtesy: https://s3.amazonaws.com/pcalc/iphone/im 1

7. Dropbox

The XS and XS Max iPhone comes with a clean and secure option for storing and syncing your files. With this app, you can easily create folders, documents and also upload photos. It also allows you to share files with other users. It still allows you to make use of the files offline.

8. Dropbox Paper

This Dropbox Paper brings the document collaboration feature to the Smartphone. It is a much better alternative to Google Docs. The app is fully designed to make it easy to embed tables and media. This app allows one to leave comments, tag your collaborators and even to track to do from all documents in one single view.

9. Pocket

This app allows you to download any article that you come across and want to read. The pocket in the iPhone XS and XS Max stripe all formatting from the downloaded material while keeping only the text and images. This makes it more beautiful to read. The iPhone screen also makes it fun to read with.

10. Just Press Record

This Just Press Record app comes with a new design in the iPhone XS and XS Max. It is so simple to make use of as it requires you to open the app, tap on the button to start recording. You can even start recording by using your Apple Watch.

Chapter Five: Exploring Other Interesting Features of the iPhone X and Fundamental How To(s)

The iPhone XS is an immediate substitution for the 2017 iPhone X, laced with a 5.8in screen costing £999. The iPhone XS Max is a bigger, costlier phone, at £1,099 and with a 6.5in screen, the largest for any iPhone.

Generally, the iPhone XS and the XS Max are a similar phone, just that one of them is bigger. This itself is fascinating, in respect to the fact that, in earlier years, the iPhone 6 Plus and 6s Plus had cameras with optical picture adjustment, while the 7 Plus and 8 Plus had dual cameras. The Max holds no such preferred standpoint this year - instead, the two phones share insignificant changes and important upgrades

Both, for example, come with enhanced stereo speakers that offer additional clarity and a more articulated feeling of party hall and room environment. This is particularly evident when you're watching motion pictures or unrecorded music recordings - it doesn't precisely feel like you're in the middle of the activity; however, the experience comes nearer to you when you use this IPhone.

The glass covering the front and back of these new iPhones is harder this year as well, and that is something worth being appreciative for. Extensively superior, the XS and XS Max prevent water and other deposits from entering the phone. This means that this phone can be left in water as deep as seven feet for up to 30 minutes.

The two phones likewise have eSIMs packaged inside, but it'll be a while before you can make use of them. The idea here is to offer dual SIM bolster - so you can add various lines of services to a single device.

The iPhone XS has precisely the same inch Super Retina OLED screen as the iPhone X, and the XS max body is likewise pretty much a dead ringer as the former. This way you will have the capacity to tell an iPhone XS from its predecessor.

Apple has also come up with the HDR10, and Dolby Vision feature into the XS line, yet even without it, films and YouTube basically sparkle on this screen. Apple's True Tone display tech has exactly the intended effect as well, modestly changing the screen's shading temperature. Certain applications and sites are interested in making use of this additional screen space as well. Apple's kind of software, like Mail and Messages, likewise flaunt additional sheets of data when you hold the Max sideways.

The means that in the new iPhone XS and XS Max, there is an additional space on the screen without applications and services using it.

On the other hand, considering to what extent it took some application engineers to refresh their product for the iPhone X's screen, we might be in for a pause. Individuals have generally expected more from huge screens on smartphones, and it's dependent upon Apple - and its engineer network - to make significant use of this additional opportunity.

Honestly, becoming acclimatized to the XS and XS Max's all-screen outline and interface can take some time. Without a conventional home button, you'll need to swipe up from the base of the display to return to your home screen.

Performing multiple tasks happens not with a dual tap of a button but rather with a swipe-and-hold from around the screen's port. Since there's no unique mark sensor, you'll be making use of your face to open your XS starting now and into the foreseeable future.

For one thing, Apple made it simpler to force close applications; now all that is necessary is a fast swipe up on an application card rather than the marginally longer press-hold-and-swipe from iOS 11.

That probably won't seem like a significant change, yet it is worth being appreciative for.

What's more, you can throw together some Siri Shortcuts - macros that let you string together activities from various applications and services to deal with dangerous undertakings - even though this procedure is marginally less demanding on the Max's greater display.

Execution is only a little smoother as well, on account of the new A12 Bionic chipset packaged inside the two phones.

Applications are easily and quickly launched here compared to the iPhone X, and as a rule, bouncing all through running applications in performing different tasks is speedier also.

With regards to this kind of common usage, the progressions are discernible, however not drastically so - that is, except if you're coming from a more seasoned iPhone. All things being equal, you're in for a big ride.

Apple claims that the A12's two elite centers are up to 15 percent quicker than the CPU centers in previous iPhones.

Apple appeared to pay more attention to its quad-center

designs processor, which can accelerate up to 50 percent quicker - which clarifies why graphically good games like Fortnite keep running with less disturbance. As games and graphically great applications turn out to be universal, the upsides of Apple's chipset plan choices ought to end up self-evident.

Chapter Six: Mastering Photography on the iPhone XS and XS Max

As previously mentioned, the iPhone XS and XS Max share another enhanced feature of the dual camera. The 12-megapixel fax sensor behind the camera bump is equivalent to the iPhone XS Max. As a general rule, you'll be shooting with the new 12-megapixel wide-point sensor. This is indeed super amazing.

This new wide-point sensor is manually bigger than the previous ones, as are the light-buttoning pixels specking it (1.4μm versus 1.22μm in the iPhone X). Put another way, the XS's primary camera is presently much better at buttoning light. Furthermore, since there are more center pixels spread over the sensor, the XS and XS Max are surprisingly fast at locking center around a focus.

Indeed, the new iPhone photography is amazing, thanks in substantial part to the A12's Neural Engine helping the camera make sense of what's in the forefront and back.

It is not yet flawless - thoro arc still occasions when the camera lets parts of a subject move away from plain sight.

Another component called Depth Control gives you a chance to change the depth of a photograph after you've effectively taken it, and it's been helpful for getting picture shots looking perfectly nice.

Also, since this component runs totally in software, you can utilize it to gussy-up your selfies as well. That is particularly useful when you consider that Apple didn't change the 7-megapixel front camera from the previous model.

The iPhone's camera has vastly improved.

With Depth Control, you can change the quality of the obscure behind your focus even in the wake of snapping a Portrait Mode shot.

The default "gap" setting is f/4.5; you can go as wide as f/1.4 (more bokeh) or as limited as f/16 (no bokeh). The Galaxy S9+, Note 9 and Note 8 all have a comparable shot. However, Apple's execution enables you to change Portrait Lighting afterward, as well.

Here's How:

Open the photograph, and you'll see the adjust option. The photograph most likely should be taken in Portrait Mode for Depth Control to be accessible.

Click on Edit.

Make use of the slider to control the introduced gap and quality of the bokeh impact. Slide for a wider opening (smaller f-stop) or left for a smaller gap (bigger f-stop). If you like, you can likewise use this medium to modify lighting conditions with the transparency signs

Click on the done button when finished. You can return to the first or make assist changes by altering again later.

Setting up and Configuration Your iPhone XS and XS Max.

https://www.cnet.com/how-to/things-to-se 1

Doubtlessly, the iPhone XS is to a great degree not the same as past iPhones. Regardless of this, the iPhone setup process hasn't changed much. Notwithstanding, while you may end up using it the way you were used to, there are still a lot of easily overlooked details you truly should look into before you start up your new iPhone.

Here is how to set up your new iPhone XS the correct way.

With iPhone XS, you'll have the capacity to exploit Apple's Automatic Setup. In case you're coming from a more established iPhone without Face ID, you will find that Touch ID is no more. That implies you'll just need to spare one face, rather than a few fingers.

In case you're a serial upgrader, and you're originating from a year-old iPhone X, less has changed. In any case, despite everything, you'll have to update equally.

Re-establishing from a Backup of Your Old iPhone

It's doubtless that you'll be re-establishing your new iPhone from a backup of your current iPhone. To do this, you just need to complete two things:

1. Ensure you have a state-of-the-art backup.

2. Make use of Apple's new Automatic Setup highlight to kick you off.

http://osxdaily.com/2017/12/31/how-acces 1 No. 1 is as straightforward as making a beeline for the iCloud settings on your iPhone and watching that there is an ongoing programmed backup. If not, simply complete one manually. From the Settings > Your Name > iCloud > iCloud Backup and tap Back Up Now. Hold up until the point that it wraps up.

Programmed Setup for iPhone XS

With respect to No. 2, Automatic Setup gives you a chance to duplicate your Apple ID and home Wi-Fi settings from another device, just by uniting them closely.

If your old iPhone (or iPad) is now running on iOS 11 or iOS 12, just put the devices side by side. This makes the underlying iPhone setup much smoother.

iPhone Setup: The Fundamentals

Re-download just the applications you require – this one is an absolute necessity. Most often users have numerous applications on their iPhones that they don't make use of. This is the most important reason, to complete a perfect setup, to be completely done.

Take advantage of the App Store application and ensure you're marked into your Apple account. Tap the little symbol of a head on the updates board to see which account you're marked in as.

How to Use Siri on iPhone XS and XS Max Like a Pro.

Consequently, Apple wouldn't dump the Home button and the majority of its highlights without outlining another edge similarly as good as the other.

With the iPhone X a year ago, this new edge came in three sections — Face ID, signals, and the Side button. These three pieces cooperate to guarantee that a buttonless iPhone display wouldn't forfeit ease of use.

To express the self-evident, you need not bother with any button to get to Siri, on any iPhone. "Hey, Siri," takes care of that. Obviously, if you don't want to say, "Hello Siri," in case anyone can hear, you'll need to know how to activate Siri quietly.

Fortunately, it's no more complex than it was on your old iPhone. Just press and hang as an afterthought button, which is once in a while referred to as the power button or rest/wake button. Following a brief moment, Siri will pop straight up

There are a few different ways you can initiate Siri on your new iPhone.

- Enact Siri on iPhone Using Your Voice

- To enact Siri making use of your voice, ensure you have set up Hey Siri on your device. If you haven't done it, open Settings →Siri and Search → Listen for Siri and set it up.

- When you have launched it, simply say, "Hello Siri" to raise the virtual right hand.

- Initiate Siri Using Side Button

- To initiate Siri manually on your iPhone X arrangement, just press and hold the side button for a few seconds.

- Exit Siri

To exit Siri, simply swipe up from the base of the display or simply press the side button to go to the home screen.

By moving the virtual right hand to the ON/OFF button,

Apple has made the side button more useful and busier. Aside from setting off the individual aide, the power button is additionally used to give you a chance to rest/wake your device, call crisis benefits, and even incidentally handicap Touch ID.

Tips to Help Beginner and Expert iPhone Users

Apple makes use of the iOS operating system. This iOS operating system powers the iPhone and iPad, as macOS on the Mac, and Windows on the PC. Apple improves on this operating system frequently, with one unrestricted refresh made accessible for all iOS users sometime in September or October every year.

What's more, iOS gets occasional smaller updates that squash bugs, fix up security gaps, and sometimes include new highlights. While the iPhone and iPad have their differences, most of the working edge is identical on both.

Setting Up Your iPhone

The basic thing to do after the purchase of the phone is to set up your new device. This implies turning it on and starting it. On an iPhone, you'll discover the power button on the right slde of the device, on an iPad you'll discover the power button on the best edge.

It is required that iPhone users enter a SIM card, by making use of the included SIM release instrument on the entryway. If you haven't included a SIM card, you won't have the ability to start-up your iPhone and continue with setup.

Setting Up Your Device

When you first turn on your device, you'll have to choose your preferred language or according to your area location.

You'll then click the Quick Start option to switch your settings from an initial iOS device. If you have a pre-iOS device, you can adhere to the on-screen guidelines; generally hit Set up manually and select a Wi-Fi system to associate it with.

Once your device is launched online, you'll then have to wait for the actuation procedure to complete loading before starting the phone

Face ID: If you have an iPhone X, you'll be requested to examine your face so that your iPhone can remember you.

Touch ID: Apple's unique finger impression. This is the need to enter your password or Apple ID secret key without fail.

Password: Even if you Make use of Face ID or Touch ID, you'll require a decent old password. Six digits is truly secure; however, you can create more difficult passwords by making use of the Passcode Options button.

Setting off a new note

Next, you'll be welcomed to either set your device up as another iPhone or iPad or re-establish from past backups.

If you have a more secure device that you wish to associate to your Mac or PC, release iTunes, and make a backup. You will then be able to pick Restore from iTunes Backup on your new device, connect it to your PC, and pick the necessary backup. The majority of your applications, individual information, contacts, and more will be moved.

If you have an iCloud backup, you can re-establish from iCloud Backup; however, since this uses the web, it will take more time. You would most likely pick up this choice if you lost your last device.

If this is the first iOS device you are using, you can either Set Up as New iPhone/iPad or Move Data from Android. Switching from Android to iPhone is really simple, since Apple currently provides an application that generally robotizes the procedure.

Creating Your Apple ID and Login

It is now expected that you have backed up or set up your new devices, the last step is to login with your Apple ID. You can create one if you don't have one by tapping the "Don't have an Apple ID button."

Apple currently uses the two-factor verification (2FA). 2FA uses two snippets of data to check your identity: something you know, and something you have on you. So, when you login, you'll have to enter your secret word initially, and after that, a code will be sent to you to confirm that it is you who initiated the login attempt.

While it might appear to be a big task to set up 2FA and input codes, the reason is to secure your account and guarantee that you alone have access to the phone.

Completing Touches

You'll have to acknowledge Apple's terms and conditions (no, you don't need to read everything), and you'll be given a chance to launch the accompanying services:

Area Services: Allows applications and different services to recognize your area. However, it must be by the express permission of you.

Apple Pay: If your local bank or country supports Apple Pay, you can include a credit or platinum card and pay for things anywhere simply by moving your Phone over the terminal. It is also helpful if you manage your wallet, and also for sending money to friends.

Siri: Apple's savvy aide does much more than just handle voice commands. You should likewise launch this.

iPhone Analytics: Sends user information from your device to Apple so that they can use it to improve on how they serve you and how you make use of your iPhone — Apple demands this information is gathered through their user agreement. So, it's dependent upon the user to agree altogether.

Application Analytics: Same arrangement as above however with outsider application designers.

Genuine Tone Display: When you launch True Tone innovation, it adjusts the whites on-screen with the temperature of the whites on the earth. Reduces eye strain in case you're gazing at a screen for quite a while.

Display Zoom: If you experience some difficulty in seeing smaller screens, you will need to launch the "Zoomed" mode to enlarge the screen.

Tips and New Easy Shortcuts and Tricks on How to Use the iPhone XS

The new improved iPhone XS and iPhone XS Max have changed the face of mobile experience. From the new improved face ID to the all-new Memoji, the iPhone XS and XS Max stand out as the best this year. Upgrading from an older iPhone or moving over from Android might be a bit difficult on this new iPhone. Here are some tips that you need to know for your iPhone XS and XS Max as you get started.

1. Set Up Face ID

As Apple finally kills the home button on the two new iPhones, iPhone XS and XS Max, the easiest way you can now unlock the phone is through the use of the Face ID. To set this up, you need to be in a place fully illuminated. And then Go to *Settings > then to Face ID & Passcode.* If you have already set up a passcode, you'll be prompted to enter it. Otherwise, you'll need to set one up to continue. Tap *Setup Face ID.*

You will also need to scan your face at least twice to complete the process. You may need to move the phone further down your face if it does not open.

Also, you can use the alternative setting by going to *Settings > Face ID and Passcode > Set up an Alternate Appearance.*

2. Take A Screenshot

Taking a screenshot is now much easier. If you were used to tapping the home button with the power button, you will now need to tap the power and volume up button simultaneously. This will bring out a small image of the screenshot, which will appear at the lower left of the screen. You can now tap the image if you wish to make any edits. Otherwise, the image will be saved to your phone.

3. Set Up Apple Pay

You can leave your wallet at home and make payments with Apple pay. With this, you can easily make payments with your phone. To set-up Apple Pay. You will need to add a card to the wallet. This can be done by going to *Settings > Wallet & Apple Pay > Add Card.* After this, you may need to visit or contact your bank to verify the card before you can use it.

Once set up, just double *https://fr-vision.com/detail/10-tips-tri 1* tap the lock button on the right side of the phone. If the phone has a Face ID set up the phone will scan your face to approve the purchase. If not, you'll need to enter your passcode.

4. Portrait and Depth Control

The iPhone XS and XS Max both come with portrait mode on the front and rear cameras. To activate it, open the camera and swipe the menu left until you see the portrait. There you will see a carousel with different lighting options. This includes the contour light, natural light stage light, and much more. Click on the desired option and press the shutter button.

5. Depth Control

This is a very new feature in the iPhone series. Depth control helps you to control the amount of blur in the background after a picture has been taken. To make use of this, simply select the desired photo that you want to use and tap edit; the Depth Control slider will appear automatically on the edit screen.

6. Turning off your Phone

In the older version of the iPhone series, the button on the right side of the device was usually the power button. But in this new series, the button is now the lock button. To turn off the phone, simply tap and hold the volume button along with the lock button together. Give it a few seconds and the slider will appear to turn your iPhone off.

7. Creating A Memoji

The last Apple product came with Amimoji for the iPhone X. this year the iPhone XS and XS Max comes with an improved feature known as Memoji. This is run in iOS 12. Memojis are animated avatars that you have the ability to control with your facial movement.

To create a Memoji, click open on the Message app and then tap on the App drawer. Select the Animoji (monkey) icon and swipe right until you see the New Memoji icon (+). Now you can easily customize your Memoji to whatever you want. When you're through, simply tap the done button at the top of the screen to save your Memoji.

8. Switch between Apps on iPhone XS?

Initially, on iPhone, you had to call on the fast app switcher to swipe back and forth between apps. With the new iPhone XS and XS Max, you can do it even faster. You can accurately just swipe. There may be a few older games or apps that might cause some difficulty, but most of the time it'll work.

This is how to do it.

1. You need to place your finger on the gesture area at the bottom of the iPhone XS or XS Max display.

2. Swipe it from left to right so as to get back to the app you were at previously.

3. Then swipe your finger from the right side to the left to return to the next app.

Note, if you get interrupted or stop at any time, the last app you were on before the interruption becomes the most recent app. Therefore, you can only swipe back from it, not forward anymore.

When you wish to quickly swipe through multiple applications on the iPhone XS or XS Max, you can also do that too. You cannot use the 3D Touch or double-click the Home button to start up the multitasking interface or also the fast app switcher.

1. Tap your finger once on the gesture area, which is at the very bottom of the iPhone XS and XS Max display.

2. Swipe up slightly. (Just place your finger on the phone screen until you get a short way up, then remove the finger)

It may take some practice to get it right. After some time, though, the gesture becomes fast.

9. Quit (Kill) Applications on iPhone XS and XS Max?

Killing an app in the iPhone XS and XS Max is so easy to do. You can do this so easy with just a simple swipe up in the app switcher tray.

1. Tap your finger on the gesture area at the bottom of the iPhone XS and XS Max display.

2. Swipe up a little.

3. Then pause. Leave your finger there, don't remove it immediately (that will take you Home). Just Pause.

4. Lift your finger.

5. Swipe up on an app card. Then it's gone.

Know that you can kill as many apps as you wish to. iOS is there to manage these apps for you. You should kill them if they really need to (looking at you, battery draining Twitter, Snapchat, Facebook, and Pokémon Go!)

10. Reachability Mode on iPhone XS and XS Max

Unlike the other gestures, you do need to set it up first.

1. Place and tap on **General**.

2. Then on **Accessibility**.

3. Click on **Reachability** to **On**.

Once set up:

Tap your finger on the gesture area at the bottom of the iPhone XS and XS Max display.

Then Swipe down.

You may even swipe down from the top right of Reachability to access Control Center.

How do you access the Control Center on iPhone XS and XS Max?

Since a swipe up now calls up the multitasking fast app switcher, Control Center is now invoked from the bottom to the top. It now means that moving Control Center to the top, Notification Center has to learn to share. Therefore, Notification Center is now limited to swiping down from the top left or from the TrueDepth camera module found in the center.

Place your finger to the right at the "horn" (where the battery indicator and cell signal is located).

Then swipe down.

Next, you can swipe down from the top right of the Reachability to access the Control Center

11.Get Benevolent with Siri

Concurrently, there is no Home button on the new iPhone XS or iPhone XS Max obviously, so to release the Siri aide you have to press and hold the Side button (on the right). On the other hand, you can simply say, "Hello Siri," when your Phone is in a silent listening area. To change this conduct, go to Siri and Search inside Settings, then kill the Listen for, "Hello Siri," flip switch.

12.Measure in increased reality

New with iOS 12 is a Measure application that exploits the iPhone's increased reality capacities. When using the plus button to check focus in three-dimensional space, you can gauge single lines or square shapes, then tap any sign to duplicate it. The shade button gives you a chance to take a photograph of your choice.

13.Check your screen time

The iOS 12 feature is centered on ensuring that your phone doesn't give you underlying stress by overusing the phone, and to that end you'll see another Screen Time passage within the Settings application. Head into the Screen Time service to discover how regularly you're using your device, and which applications are taking up the majority of your time.

14.Enact the True Tone display

Like the iPhone X before them, the iPhone XS and the iPhone XS Max have a True Tone display that changes the splendor and warmth of the screen to coordinate the encompassing light and lessen the strain on your eyes. You may have launched it when you set up your iPhone, yet you can check by heading to the iOS Settings application when tapping Display and Brightness.

15.Quieten down notifications

You can take more control over notifications with iOS 12 – you can caution from certain applications, which implies they go straight to the Notification Center without showing up on the boot screen, flying up as a flag, or making a sound. To do this, swipe left on the notice as it touches base on the screen, then tap Manage, then tap Deliver Quietly.

16.Zoom into Scenes

The iPhone XR is almost in the same class as the iPhone XS and iPhone XS Max, significantly less expensive, though brighter. Hence, you can exploit one of the advantages of the costlier phones by making use of optical zoom. From the camera button screen, with Photo chose, tap on the 1x button to hop to a 2x optical zoom view.

17.Set up the Second Face

Unlike other versions of the iPhone, your face serves as the optimal security passcode instead of a fingerprint on the new iPhone version. New in iOS 12 on the iPhone XS and iPhone XS Max (and iPhone X and iPhone XR) is the capacity to set a second face for opening your phone, either for your loved ones or accomplice, by using the Face ID and Passcode in Settings – tap Set Up an Alternative Appearance to begin.

Made in the USA
Lexington, KY
11 March 2019